ISBN: 978-0-557-03024-8

Published by lulu.com

Dr. Fred B. Wood, III & Andrea Bowe

Csiri.org:

Climate Change and Advanced Electromagnetics
Prepared for ISSS (International Society for

Systems Science) Conference,

Toronto, Canada. July, 2000.

 □   **Climate and Advanced Electromagnetics**

A Start of a Proto-Encyclopedia of Systems Theory
Concepts Needed for Relating Climate Cycles to the
Survival of Human Civilization.

                        Fred B. Wood, Sr., B.S.,
M.S., Ph.D. (Electrical Engineering) (UC
Berkeley)□Non-Profit: Computer Social Impact

Research Institute, Inc., a 501 (c)(3) since 1978.

"To Share and Care For All" and *"Law of The Seventh Generation"*

by Andrea Bowe,

Director, CSIRI, 2007-2008.

Candidate, Ph.D. in Education and Leadership, Self-Designed, Walden University, June, 2005-Present.

Master's in Education in Instructional Technology, Grand Canyon University, May, 2005, (gpa 4.0).

Independent Producer, communitytv.org, SC CA.□

2 year Weekly Series: Energy Crisis: Fact or Fiction? (2001-2003).

Honorable Mention: EarthVision Contest for 78

minute documentary shown New-Year's Day, 2003 @
9:30PM on Cable CH 27□"Is Art and Music in Santa
Cruz an Endangered Species?" submitted under the
category Endangered species and Habitats.

Bachelor of Science, Liberal Studies, 1974,
OSU (Oregon State University) graduated with High
Scholarship (gpa 3.69)

Nuclear Reactor Technology as a minor with a
General Engineering & Psychology Major @ Oregon
State Univ. □Honors Student(OSU).

  Electrical Engineering and Communication Arts,
including several roles in on-stage productions by
the Del Rey Players @ Loyola University, LA (gpa
3.96)

National Honor Society Scholarship recipient for
two years: 1969-1971.□

Lifetime Member. (NHS: National Honor Society)□

Honors@Entrance, Loyola.

1969 Valedictorian, Notre Dame Academy Girls' High

☐

Early Entrance to Loyola University between junior and senior years in HS.

Lifetime Member of Ca Scholarship Federation (CSF).

Member, BEA (Broadcast Education Association), 2006-2008.

Piano Teacher, Sherwood Music School Chicago, IL. Student in correspondence course from 1960-69.

Composer/Musician/Poet/Web Page Design/Editor/E-Publisher, 17 books on lulu.com

☐Master's in Education in Instructional Technology, GCU online, May 2005 (gpa 4.0)

Computer Social Impact

Research Institute, Inc. (non-profit since 1978)

Email Andi Bowe @ abowe@csiri.org

☐

Twenty-eight systems-component areas of science, electrical engineering, bio-science, and spirituality are listed which require research to facilitate the glacial cycle being distorted by the CO/2 added to our atmosphere by the burning of fossil fuels. A sampling of Fig. 1, Fig. 5 (Roots of Differential Equations of Electricity and Magnetism) and Fig. 11 are displayed below to show the path of research used here.

This report also constitutes a base of 28 component areas for monitoring the uses of the energy available in the Quantum Vacuum for four purposes: (Connected with a proposal of Ms. Andrea Bowe of Santa Cruz, California, for a Spiritual UN):

(1) Developing an understanding of "Consciousness" related to the Quantum Vacuum Holofield, □

(2) Monitoring the Weapons being developed for World War III, to restrict the STEALING of Quantum Vacuum Energy from the indigenous peoples' healing resources, □

(3) To develop a dialog between the major political forces on our planet on the proper uses of the three classes of electromagnetic waves for the benefit of humankind and the planet, and □

(4) Establishing an unclassified Manhattan-type

Project to design and produce "over-unity " electric generators so that every country in the world could produce enough electricity for it's own use, without having to go to war for energy sources such as clear cutting of trees, coal mining, and drilling for oil.

(5) Development of legal protection for trees that may be the carrier devices (like telephone poles holding wires) for maintaining the integrity of the holofield.

There may be an opportunity to help some develop an understanding of the three major types of electromagnetic waves: Hertzian, Biological, and Information Carrying Waves like Tachyon and Torsion waves described by engineers or radial waves as envisioned by Teilhard de Chardin to represent

Love. □

This paper includes two abstracts:

First Abstract: Prepared in November 1999, and a
Second Abstract: Evolved in steps as the use of
systems theory uncovered more and more elements of
the system. □ □

FIRST ABSTRACT □

WOOD, F. B., SR, and Bowe, A. G.

CLIMATE CHANGE AND ADVANCED ELECTROMAGNETICS:FUTURE

Computer Social Impact Research Institute, Inc.
(non-profit) csiri.org

Recent apparent evidence of global warming once
again highlights the possibility of debate over the
nature and extent of climate change, and what we
might do to stabilize the global climate should we

desire to do so. Using my green cube overview of
systems analysis, we determined that the analysis
of the climate change problem involves many
different fields of science.

☐Advanced Electromagnetics might be part of the
solution, because of its potential to contribute to
breakthrough developments in energy production and
distribution, transportation and other energy-
intensive sectors that are now producing high
levels of so-called greenhouse gases. However, this
same advanced EM may have been already weaponized
to some degree, thus impeding its use for peaceful,
civilian purposes. Overcoming this obstacle
probably requires: interdisciplinary systems
research, dialog sessions, diplomatic initiatives
and creating win-win strategies.

SECOND ABSTRACT

As the work proceeded, we found more and more
connections between the problem of atmospheric

warming and other areas such as soil nutrition, biological systems, and questions of ethics leading to more detailed investigation of the role of spirituality in the survival of civilization.

We propose Hypotheses A and B and C as the most significant results coming out of this study:

Hypothesis A: European and American Electrodynamics and Electrical Engineering have had a serious defect since 1872, 128 years ago when Heaviside (Hunt, 1991) simplified James Clerk Maxwell's equations for electromagnetic radiation by leaving out the scalar potential terms. The simplified Maxwell's Equations were adequate enough for designing the microwave RADAR systems of World War II (1939-1945), but not comprehensive enough for designing the type of RADAR anticipated for World War III. (See work of Thomas Bearden in "Fer-De-Lance A Study of Soviet Scalar Electromagnetic Weapons.")

Hypothesis B: It is estimated that, by about 1952
(48 years ago), Joseph Stalin asked the top Soviet
scientists to organize a project of the order of
magnitude of the American Manhattan Project of
World War II to explore the corrections needed to
the textbook versions of Maxwell's Equations and
the weapons designs of Nicola Tesla (1904-1910)
that would make the Soviet Union invincible in
World War III. There is evidence that the secret
Soviet weapons developments have achieved
considerable success. Brezhnev, Krushchev, and
Gromyko each in turn made public appeals for the
United Nations to establish a committee to
supervise the development of new weapons of mass
destruction. Our country, the United States, did
not cooperate, saying that we thought the Soviet
proposals were a trick to get us to reveal our
future weapons. It is important that we at least
belatedly take the initiative in negotiating for
discussions of the new weapons. Furthermore, we are

learning that part of our human brain memory system is outside the skull in the quantum vacuum holofield where both American and Russian brains are sharing the same holofield with an elementary security code provided by Galois field multiplier polynomials for coding.

Hypothesis C: There is a severe problem with capitalist economic systems in that they try to prevent their citizens from understanding how the economic system works, so that members of the power-elite can manipulate energy policies to conceal national policies. For example, over-unity engines and electric generators could be used to solve some energy problems. The suppression of the development of over-unity engines has greatly increased the pollution of the environment, and led a growing population to think it is necessary to go to war to protect our oil deposits. This means that, if the United States supported the development of over-unity engines, instead of delaying it, people could get all the electric power they wanted, so there would be no power drive for people to go to war for oil or other energy deposits, because everyone could get the power they wanted. ☐They could get it by making use of over-unity engines and generators which do not pollute the environment. The drive for more oil, gas, and coal may have stimulated the drive for World War I and World War II, which led to 320,000 casualties in WW I ☐and 1,078,000 casualties in WWII. The speedy development and distribution of overunity electric motors and generators is the simplest way to move toward peace. ☐   ☐

TABLE OF CONTENTS

DESCRIPTION OF COMPONENTS

Components of Earth Planet that form the principle elements in a complex natural system that are here represented by picto-grams or graphs of numerical data, to be followed by more detailed descriptions.

Fig. 1. Curves of Rise of Atmospheric Carbon Dioxide and Global Surface Temperatures During Last 140 Years (Karl, 1999) (Schneider, 1989).

Fig. 2. Last 4.6 Billion Years: Estimate of Cold Cycles, Ice Eras, Ice Epochs and Ice-Age Cycles.(Wood, 1998)

Fig. 3A. 2.4 Million Year Ice Epoch. (Kukla, 1992)

Fig. 3B. Last Ice-Age Cycle.(Kukla, 1992)

Fig. 4. Typical Glacial Climate cycle. (How much is base cycle in Nature; and how much is the influence of Humankind ?) (Felix, 1997) (Hamaker, 1982) (Sitchin, 1985) (Tompkins, 1989)

Fig. 5. The Green Cube - Template - For Library of
Evolution, Containing dimensions of x: Class of
activity; Y: Levels of Phenomena; Z Time Stages of
Evolution and Cycles. (Pond, 1990, 1996)
(Schneider, 1994) (Wood, M., 1987) (Wood, F.B.,
1998)

Fig. 6. Proposed Metcalf Energy Center, San Jose,
Peak Processing; 600 Megawatts derived from natural
gas turbine plus 600 megawatts energy abstracted
from the the Quantum Vacuum for a total of 1200
Megawatts energy peak processing. The system will
then use the old 1890 designs for generating
electric power which can at most only utilize half
of the electric power generated, so they will
radiate up to 600 megawatts of electric energy into
outer space of our universe. Nicola Tesla thought
that it was crazy for American power companies to
throw away half of the electric power generated. J.
P. Morgan of Wall Street arranged to call the loans
to Tesla Electric Co., putting Tesla into

bankruptcy. Free room and board was made available to Tesla at the New Yorker Hotel for the rest of his normal working life. (Manning,1996) (Bearden, T., 2000, see Appendix B)

Fig. 7. Vortexes or Spirals of Energy used to be included in engineering textbooks of the 1890 era. After 1904, Tesla claimed that there were more kinds of electromagnetic waves than were described in text books. Natural healers use vortexes in their description of the coupling of bioenergy to human system.

Fig. 8. In the mid 1980's structures similar to some of Tesla's sketches of the 1904-1910 era of new weapons and defenses against them were seen at missile testing ranges in Central Asia through photographs made by high altitude spy planes.(Bearden, 1986) (Hunt, 1991)

Fig. 9. Woodpecker radio beams intersect over North America to form Interfereometer Class of Radar

Operations. The energy in the radar beams forms standing waves which interact to form a coordinate system over the target area for identifying airplanes or missles to be shot down. (Schwartz,1998) (Weinberg, 1996) (Wood, 1998)

Fig. 10. Test of Proposed World War III Scalar Electromagnetic Weapon to disable an Airplane over the Ocean without Killing anyone. (Bearden,1991, Gravitobiology)

Fig. 11. N-1 Homopolar Generator Over Unity Electric Generator of 300 Kilowatt Maximum Output, Input 100 KW, Average Output 184 KW. This Model used to demonstrate the principles. The added 84 KW output came from the Quantum Vacuum Flux or the Primordial Field of the Universe - a sea of 'free energy' which permeates all. - described by Bruce DePalma as the Primordial Field. This model gave overunity power ratio of 1.84 in power range of 300 kilowatts. An earlier model at 7.56 kw gave a ratio of 28.2 in 1980 reported by De Palma in article by Robert Kincheloe in 1986 in The Free Energy Device Handbook,(1994) Stelle, IL: Adventures Unlimited Press. A number of other free-energy devices are reviewed in this Handbook.(Childress, 1994) (Manning, 1996) (Quigg, 1983, 1997)

Fig. 12. Example of Division by Code 1 Polynomial in a Galois Field of binary numbers that are modulating the Holofield.

Fig. 13. God-Trees (Redwood Trees, Cedars of Lebanon, and others) whose electrical aura supports the Holofields auxiliary memory shared by living beings.

Fig. 14. The Holofield and Noosphere of the Quantum Vacuum Flux with different carrier frequencies assigned to different species of living animals and plants which are modulated by galois fields with different code polynomials assigned to different individuals and groups of individuals. (Sharpe,

1993)

Fig. 15. The Heart Intelligence of Human Beings and
of other animals maintains communication with the
Holofield on principles and sends requests for
details to its brain intelligence with comparisons
with the highest ideals stored in the holofield by
previous generations. These need to be worked out
in finer detail using the block diagrams of F.
Wayne Kraft. (Childre, 1999) (Gordon, 1999) (Kraft,
1983) (Starhawk 1993)

Fig. 16. The Human BRAIN has communication
channels, storage cells, and logic structures with
which to work out the details of the principles
generated by the Heart Intelligence. I don't find
any consensus on how the three kinds of
electromagnetic waves have particular roles in the
functioning of human consciousness. I am trying to
determine the role of the three types of
electromagnetic waves that are derived from the

three roots of the partial differential equations of electricity and magnetism. Russian scientists have derived three types special electromagnetic waves from different configurations of spinning charges. They report that their "torsion" waves may have an important role in consciousness. (Ornstein, 1997)

Fig. 17. The CHURCHES have a role in helping people develop their consciousness to be able to follow an evolutionary path by connecting to information on ethics stored by previous generations in the holofield.. (Fagg, 1999)

Fig. 18. EARTHLING here represents Homo Sapiens on Planet Earth. Examining the " Earth Chronicles" series of books by Zecharia Sitchin, I find that Sitchin's interpretation of ancient tablets point to a planet, Nibiru,where civilization had reached a point more advanced than our civilization, having developed satisfactory interplanetary space ships.

They planned to make gold foil sheets to reflect
some of the sunlight away to stabilize the rising
planet surface temperature, like in Fig. 1 on Earth
currently. Nibiru people couldn't find enough gold
on their planet, so they traveled in their space
ships to EARTH to mine gold and transport it back
to planet Nibiru. (Brennan, 1987, 1993) (Kessler,
1991) (La Violette, 1994, 1995)

Fig. 19. I show an ET also communicating with the
HOLOFIELD.(Clow,1995)

Fig. 20. The LABYRINTH has become more popular in
American churches lately. People are improving
their health by walking slowly through the
LABYRINTH. Probably they are receiving a small,
equally distributed, dose of bioenergy (2nd order
electromagnetic) waves.

Fig. 21. Here we have Barbara Marx Hubbard's
concept of conscious evolution. (Anderson, 1993)
(Begich, 2000) (Hubbard, 1977, 1995, 1998)

Fig. 22. Here is a symbol for a cybernetics feedback prototype to help us direct any such process and be helpful in finding more beneficial changes in social structure.

Fig. 23. Here is a symbol of the Heart and Holofield Negentropy Production (Consoletti, 1998) (Peat, 1997)

Figs. 24, 25, & 26: Here we show the communication with HOLOFIELD for: □24: EARTHLING, (Johari, 1987) (Laszlo, 1999) (Targ, 1998) (Yogananda,1946) □25: GOD-HEAD, (Hurtak, 1977) □26: ET (Walker, 2000)

Fig. 27. NON-HIERARCHIAL organization like that proposed by Sir Stafford Beer.(Bohm, 1999) (Henderson, 1990) (Hock, 1999)

Fig. 28. A proposal by Andrea Bowe for a Spiritual United Nations. We have recently learned that Bishop Swig of Grace Cathedral, San Francisco, is currently organizing an United Religious Initiative

similar to the Spiritual U.N. Motto: "To share and
care for All" (Bovard, 1994-95) ☐ ☐

CONCLUSIONS

We have had glacial cycles for almost one billion
years on planet Earth. We cannot eliminate them,
but we can reduce the magnitude and rate of coming
of the glaciers. To do our best to ameliorate the
glaciers and adapt to glacial conditions, we must:
☐(1) Continue our research on glacial cycles, ☐(2)
Educate the public on these issues. ☐(3) Develop
energy sources for heating buildings and caves
during glaciation, ☐(4) Stop burning fossil fuels
to slow down the increase of carbon dioxide in the
atmosphere, ☐(5) Provide a job retraining program
for workers in the fossil fuel industries, ☐(6)
Provide an investment trading of stocks and bonds
of fossil fuel industries for treasury bonds at
equitable rates to insure continuity of pension
funds, school funds, and non-profit funds, ☐(7)
Provide for treasury bond trade with forest land

owners for the government taking over their old growth forests, ☐(8) Develop overunity engines to generate electricity, ☐(9) Develop overunity engines to provide a power source for automobiles, trains, trucks, busses, ☐(10) Develop overunity engines for airplanes, ☐(11) Also develop overunity motors for running rock grinders for remineralizing the soil, ☐(12) Plant more forests to absorb carbon dioxide, etc. ☐(13) Contract with the Chaordic Alliance (Non-Profit 501(c)(3)), Half Moon Bay, California, to have Affiliates provide periodic review of how well the above projects are progressing, ☐(14) Establish Research Construction Company to fill in any gaps in the program to withstand glaciation. ☐(15) Improve our communications capability with people on planets in other solar systems to see if they have any better ideas for dealing with the glaciation. ☐(16) Review old books like F. Creedy, The Next Step in Civilization. ☐ ☐ ☐

APPENDICES

APPENDIX A: LIST FOR CHECKING HYPOTHESES IN SCIENCE

AND

ENGINEERING.

IDENTIFICATION OF STAGE OF DEVELOPMENT OF

INCOMPLETE MANUSCRIPTS

The reason for the existence of this series of "working paper drafts" is that preliminary interdisciplinary studies of the interaction between engineering science and society have to be done by conference or by correspondence between individuals who are interested in the subject but may be geographically distant from each other. After some correspondence and exchange of notes, the material may be referred to someone in the social sciences who may be better prepared to continue the investigation of the problem or it may be revised for submission to one of the engineering or scientific journals.

Where articles may be at different stages of development, we have to pay more attention to the process of solving problems, forming and testing hypotheses in the process of scientific research. It is important that such work be properly identified so that preliminary conclusions are not confused with logical conclusions. An investigation of the social consequences of some invention or a new scientific concept might go through the following stages: ☐A: Searching for background reference material ☐B: Brief reading of background material ☐C: Study and discussion of background material ☐D: Summarizing of previous research workers' results. ☐E: Definition of the immediate problem ☐F: Tabulation of references for the specific problem ☐G: Brief study of the specific

problem □H: Formulation of preliminary hypotheses
□I: Checking of preliminary hypotheses for
agreement with known data. □J: Collection of new
data or setting up experiments □K: Checking of
hypotheses with cultural values in art, music,
poetry, ethics, religious experience,etc. □L:
Critical testing and revision of hypotheses □M:
Preliminary report writing □N: Circulation of
preliminary reports to others for criticism □O:
Revision of reports or oral presentation at
scientific societies □P: Publication of Articles
□Q: Experimental use of ideas with small groups of
specialists and/or laymen □R: Review of the value
of hypotheses as tried in practice □S: Preparation
of popularized versions for public use □T:
Preparation of more technical versions for social
science research □In some cases the next letter is
used to indicate the next version of the analysis,
even if the stage doesn't fit the above definitions
precisely. Frederick B. Wood, Ph.D. 12/24/45 □

APPENDIX B: EXTRACT FROM T. E. BEARDEN, 36 PAGE

STATEMENT:

"The Unnecessary Energy Crisis: How to Solve it
Quickly," from pages 11-16

Electrical Energy Required from Hydrocarbon Burning
Drives the Problem □The heart of the present
environmental pollution problem is the ever-
increasing need for electrical energy obtained from
burning of hydrocarbon fuels and/or nuclear power
stations. The increasing production of electrical
power to fill the rising needs, increasingly
pollutes the environment including the populace
itself (lungs, bodies, etc.). Almost every species
on earth is affected, and every year some species
become extinct as a result. Environmental pollution
includes pollution of the soil, fresh and salt
water, and the atmosphere by a variety of waste

products. Given global warming, it also includes excess heat pollution in addition to chemical and nuclear residues. Under present procedures, the electrical energy problem is exacerbated by decreasing available oil supplies, which are believed to have peaked this year, with a projected decline from now on. But really, the electrical energy problem is due to the scientific community's adamant defense and use of electrical power system models and theories that are 136 years old { } in their very foundations, riddled with errors and non sequiturs, and seriously flawed. The scientific community has not even recognized the problem, much less the solution. In fact, it does not even intend to recognize the problem, even though the basis for it has been known in particle physics for nearly 50 years. As Bunge { } put it some decades ago: ☐"...it is not usually acknowledged that electrodynamics, both classical and quantal, are in a sad state." ☐The scientific community has done little to correct that fundamental problem since Bunge made his wry statement. □Let us put it very simply: The most modern theory today is modern gauge field theory. In that theory, freedom of gauge is assumed from the getgo. Applied to electrodynamics, this means — as all electrodynamicists have assumed for the last century or longer — that the potential energy of an EM system can be freely changed at will. In other words, in theory it costs nothing at all to increase the EM energy collected in a system; this is merely "changing the voltage", which does not require power. In other words, we can "excite" the system with excess energy (actually taken from the vacuum), at will. For free. And the best science of the day agrees with that statement. It also follows that we can freely change the excitation energy again, at will. In short, we can dissipate that excess energy freely and at will. Without cost. Well, this means that we are free — by the laws of nature, physics, thermodynamics, and gauge field theory — to dissipate that free excess potential energy in an external load, thus doing "free work". Since none of the systems our energy scientists and engineers build for us are doing that, it follows a priori that the fault lies entirely in their own system design and building. It does not lie in any prohibition by nature or the laws of physics. A priori, then, the present COP<1.0 performance of our electrical power systems is a monstrosity and the direct fault of our scientists and engineers. We cannot blame the laws of nature or the laws of physics. ▯The present energy crisis then is due totally to that "conspiracy of ignorance" we referred to, that is maintained by the scientific community, and that has been maintained by it for more than 100 years. ▯This is the real situation that the environmentalists must become aware of, if they are to see the correct path into which their energies and efforts should be directed — to solve both the energy crisis and the problem of gigantic pollution of the biosphere. □Outside Intervention Must Forcibly Move Energy Science Forward Unless outside intervention occurs forcibly, the scientific community's lock-up of research funds for "in the box" energy research may result in the economic collapse of the Western World in perhaps as little as eight years. Let us examine the gist of the problem facing us. Suppose we launch a crash program to develop, manufacture, deploy, and employ the new "vacuum powered" systems. Once the new self-powering systems are developed and ready to roll off the production lines en masse, it will require a minimum of five years worldwide to sufficiently alter the "electrical energy from oil" demand curve, so that economic collapse can be averted. In turn, this means that the new systems must be ready to roll off the manufacturing lines by the end of 2003. While this is a very tight schedule, it can be done if we move rapidly. The necessary scientific corrections along the lines indicated in this paper can be quickly applied to solve the electrical energy problem permanently and economically, given a Manhattan type project under a Presidential Decision Directive together with a Presidential declaration of a National Energy Emergency. □In a paper { } to be published in Russia in July 2000, this researcher has proposed some 15 viable methods for developing new "self-powering" systems powering themselves and their loads with energy extracted from the vacuum. Several of these systems can be developed very rapidly and easily mass produced. □A second paper { } will be published in the same proceedings, revealing the Bedini method for invoking a negative resistor inside a storage battery. The negative resistor freely extracts vacuum energy and adds it to both the battery-recharging function and the load powering function. □In Bedini's negative resistor method, decoupling (dephasing) the ion current inside the battery from the electron current between the outer circuit and the external surfaces of the battery plates, allows the battery to be charged (with increased charging energy) simultaneously as the load is powered with increased current and voltage. At my specific request, both papers were thoroughly reviewed by qualified Russian scientists, and the premises passed with flying colors. ⬜A third paper { } gives the exact giant negentropy mechanism by which the dipole extracts such enormous energy from the vacuum. We will further explain that mechanism below. ⬜Conventional Approaches: Too Little, Too Late ⬜It appears that the Environmental Community itself has finally realized that the present scientific approaches and research are simply too little and too late. Further, the conventional approaches are largely "in the box thinking" applied to an "out of the box problem." We leave it to others such as Loder { } to succinctly summarize the shortfalls of these present solutions. Loder, e.g., particularly and incisively explains how the problem with automobiles breaks down. In fact, no one single COP>1.0 approach will be all-sufficing. Several solutions, each for a different application, must be developed and deployed simultaneously. ☐As an example, it is possible to create certain dipolar phenomena in plasmas produced in special burners, such that the dipoles extract substantial excess EM energy from the vacuum and output it as ordinary excess heat well beyond what the combustion process alone will yield. Given a Manhattan type project, the inventor of that process (with already working models and rigorous measurements) could rapidly be augmented to develop a series of replacement burners (heaters) for ordinary electrical power plants to use in heating the water to make the steam for the steam turbines turning the shafts of the generators. The entire remainder of the power system, grid, etc. could be left intact. Some fuel would still be burned, but far less would be consumed in order to furnish the same required heat output. □In short, a rather dramatic reduction in power plant hydrocarbon combustion could be achieved — in the present electrical power plants with minimum modification, and in the necessary time frame — while maintaining or even increasing the electrical energy output of the power systems. We believe the inventor would fully participate in a government-backed Manhattan type energy program where a National Emergency has been declared, given a U.S. government guarantee that his process, equipment, and inventions will not be confiscated. { } □Another process capable of quick development and enormous application is the development of point contact transistors as true negative resistors {30}. □Two other processes that can be developed for massive production in less than two years are (i) the Kawai process {18}, and (ii) the magnetic Wankel process {19}. In addition, the Johnson {20} process can be developed and readied for manufacture in the same time frame, given a full-bore sophisticated laboratory team. □There are other processes { } {51} {52}which can also be developed rapidly, to provide major contributions in solving their parts of the present "electrical energy from hydrocarbon combustion" problem. □Giant Negentropy and a Great New Symmetry Principle □We now summarize some recent technical discoveries by the present author that bear directly upon the problem of extracting and using copious EM energy flows from the vacuum. Any dipole has a scalar potential between its' ends, as is well-known. Extending earlier work by Stoney { }, in 1903 Whittaker { } showed that the scalar potential decomposes into — and identically is — a harmonic set of bidirectional longitudinal EM wavepairs. Each wavepair is comprised of a longitudinal EM wave (LEMW) and its phase conjugate LEMW replica.

Hence the formation of the dipole actually initiates the ongoing production of a harmonic set of such biwaves in 4-space { }. We separate the Whittaker waves into two sets: (i) the convergent phase conjugate set, in the imaginary plane, and (ii) the divergent real wave set, in 3-space. In 4-space, the 4th dimension may be taken as -ict. The only variable in -ict is t. Hence the phase conjugate waveset in the scalar potential's decomposition is a set of harmonic EM waves converging upon the dipole in the time dimension, as a time-reversed EM energy flow structure inside the structure of time { }. Or, one can just think of the waveset as converging upon the dipole in the imaginary plane { } — a concept similar to the notion of "reactive power" in electrical engineering. The divergent real EM waveset in the scalar potential's decomposition is then a harmonic set of EM waves radiating out from the dipole in all directions at the speed of light. As can be seen, there is perfect 4-symmetry in the resulting EM energy flow, but there is broken 3-symmetry since there is no observable 3-flow EM energy input to the dipole. Our professors have taught us that output energy flow in 3-space from a source or transducer, must be accompanied by an input energy flow in 3-space. That is not true. It must be accompanied by an input energy flow, period. That input can be an energy flow in the 4th dimension, time — or we can consider it as an inflow in the imaginary plane. The flow of energy must be conserved, not the dimensions in which the flow exists. There is no requirement by nature that the inflow of energy bein the same dimension as the outflow of EM energy. Indeed, nature prefers to do it the other way! Once one unties nature's foot from the usually enforced extra condition of 3-space energy flow conservation, nature joyfully and immediately sets up a giant 4-flow conservation, ongoing, where enormous EM energy is inflowing from the imaginary plane into the source charge or dipole, and is flowing out of the source charge or dipole in 3-space, at the speed of light, and in all directions. In other words, nature then gladly gives us as much EM energy flow as we need, indefinitely — just for paying a tiny little bit initially to "make the little dipole." After that, we never have to pay anything again, and nature will happily keep on pouring out that 3-flow of EM energy for us. This is the giant negentropy mechanism I uncovered, performed in the simplest way imaginable: just make an ordinary little dipole. □We may interpret the giant negentropy mechanism in electrical engineering terms { }. The EM energy flow in the imaginary plane is just incoming "pure reactive power" in the language of electrical engineering. The outgoing EM energy flow in the real plane (3-space) is "real power" in the same language. So the dipole is continuously receiving a steady stream of reactive power, transducing it into real power, and outputting it as a continuous outflow of real EM power. Further, there is perfect 1:1 correlation between the convergent waveset in the imaginary plane and the divergent waveset in 3-space. This perfect correlation between the two sets of waves and their dynamics represents a deterministic re-ordering of a fraction of the 4-vacuum energy — a re-ordering initiated by the formation of the dipole, and spreading radially outward at the speed of light. □This clearly shows that (i) we can initiate reordering of a usable fraction of the vacuum's energy at any place, anytime, easily and cheaply (we need only to form a simple dipole), and (ii) the process continues indefinitely, so long as the dipole exists, without the operator inputting a single additional watt of power. And the greatest benefit of all is that, so long as the dipole exists and this re-ordering continues, a □copious flow of observable, usable EM energy pours from the dipole in all directions at the speed of light. This is the full solution to the first half of the energy crisis, totally solved once and for all. □Ansatz of the Major Players □Scientific Community □For the most part, the organized scientific community varies from highly resistant to openly hostile toward any mention of extracting copious EM energy from the active vacuum. The "Big Nuclear" part of the community is particularly adamant in this respect, as witness its ferocious onslaught on the fledgling and struggling cold fusion researchers — a ferocity of scientific attack seldom seen in the annals of science { } { }. The scientific community also largely suppresses { } or severely badgers scientists attempting to advance electrodynamics to a more modern model, suitable to the needs of the 21st century and the desperate need for cheap, clean, nonpolluting electrical power worldwide {12}. The community still applies classical equilibrium thermodynamics to the electrical part of all its electrical power systems, even though every EM system is inherently a system far from equilibrium with the active vacuum environment, and a different thermodynamics applies. Only if the system is specifically so designed — e.g., so that during the dissipation of its excitation energy it enforces the Lorentz symmetrical regauging condition — will the system behave as a classical equilibrium system. The thermodynamics of open dissipative systems is well-known { }. Such systems are permitted to (1) self-order, (2) self-oscillate or self-rotate, (3) output more energy than the operator inputs (the excess energy is freely received from the active environment), (4) power itself and its load simultaneously (all the energy is taken from the active environment, similar to a windmill's operation), and (5) exhibit negentropy. □That our present electrical power systems do not do these five things, even though every one of them is an

open system in violent energy exchange with the
vacuum, a priori reveals that it is the scientific
model and the engineering design that are at fault.
It is not any law of nature or principle of physics
that prevents self-powering open electrical power
systems. Instead, it is the scientific community
and its prevailing mindset against extracting and
using EM energy from the vacuum. Environmental
Community. ☐In the past the environmental community
has been overly manipulative with respect to
physics, and particularly with respect to
electrical physics. Its science advisors have come
mostly from the conservative "in the box"
scientific community. Hence the community has
failed to realize that COP>1.0 electrical power
systems are normal and permitted by the laws of
nature and the laws of physics. They have no
inkling that enormous Heaviside discovered — in the
1880s! —the unaccounted EM energy pouring from the
terminals of any battery or generator. They are

unaware that Poynting considered only the tiny
component of the energy flow that enters the
circuit. They are also unaware that, completely
unable to explain the astounding enormity of the EM
energy flow if the nondiverged (nonintercepted)
Heaviside component is accounted, Lorentz {9} just
arbitrarily used a little procedure to discard that
troublesome Heaviside "dark" (unaccounted)
component. Lorentz reasoned that, since the huge
dark energy flow component missed the circuit
entirely, it "had no physical significance." This
is like arguing that none of the wind on the ocean
has any physical significance, except for that
small portion of the wind that strikes the sail of
one's own sailboat. It ignores the obvious fact
that whole fleets of additional sailboats can also
be powered by that "physically insignificant" wind
component that misses one's own sailboat entirely.
Nonetheless, electrodynamicists continue to use
Lorentz's little discard trick, and try to call the

feeble Poynting energy flow component caught by the circuit the entire EM energy flow connected with it. This is like arguing that the component of wind hitting the sails of one's own sailboat, is the entire great wind on the ocean. ☐As a result, the environmental community has failed to grasp the real reason for the energy crisis and the increasing pollution of the biosphere. They have been deceived and manipulated into thinking that conventional organized science is giving them the very best technical advice possible on electrical power systems. The environmentalists have been and are further deceived into believing that the conventional scientific community is advocating and performing the best possible scientific studies and developments for trying to solve the energy crisis. Of major importance, the environmental community itself has been deceived as to the exact nature of the energy flow in and around a circuit, the vastness of the unaccounted energy flow (or even that any of the energy flow is deliberately unaccounted), and the fact that this present but unaccounted EM energy flow can be intercepted and captured for use in powering loads and developing self-powering systems. Worst of all, the environmental community has been deceived as to what powers every electrical load and EM circuit, and that burning all those hydrocarbons, using those nuclear fuel rods, building those dams and windmills, and putting out solar cell arrays are necessary and the best that can be done. In short, they have been smoothly diverted from solving the very problem — the problem of the increasing pollution and destruction of the biosphere — they are striving to rectify. However, judging from their continued demonstrations in the street, at least many environmentalists now suspect that much of the world's continued policy of "the rich get richer and the poor get poorer" in international trade agreements are deliberately planned and implemented to the advantage of a favored financial class and the exploitation of the poorer laboring classes in disadvantaged nations. Numbered references and the complete paper may be found at http://www.cheniere.org Bearden's page ☐

APPENDIX C: LAW OF THE SEVENTH GENERATION

"DECLARATION OF THE FOUR SACRED THINGS" ☐The earth is a living, conscious being. In company with cultures of many different times and places, we name these things as sacred: air, fire, water and earth. Whether we see them as breath, energy, blood and body of the Mother, or as the blessed gifts of a Creator, or as symbols of the interconnected systems that sustain life, we know that nothing can live without them. To call these things sacred is to say that they have a value beyond their usefulness for human ends, that they themselves become the standard by which our acts, our economics, our laws and our purposes must be judged. No one has the right to appropriate them or profit from them at the expense of others. Any government that fails to protect them forfeits its legitimacy! All people, all living things, are part of the earth life and so are sacred. No one of us stands higher or lower than any other. Only justice can assure balance: only ecological balance can sustain freedom. Only in freedom can that fifth sacred thing we call spirit flourish in its full diversity. To honor the sacred is to create conditions in which nourishment, sustenance, habitat, knowledge, freedom and beauty can thrive. To honor the sacred is to make love possible. □To this we dedicate our curiosity, our will, our courage, our silences and our voices. To this we dedicate our lives."

"Our 'guts' are not in question here. If courage were all we needed, we would not be worried. But wars are not fought just with guts, or even with weapons. They are struggles of consciousness… Only a miracle can save us. And who will make miracles for us if not ourselves?" (Starhawk, 1993, pps. 236-7)

LAW OF THE SEVENTH GENERATION:

WE, THE PEOPLE, in order to form a more perfect union dedicated to the salvation of the planet Earth (HEARTH HOME HEART) RESPECTFULLY REQUEST ALL HUMANITY to join in a cosmic purpose, united in solidarity with all peoples and all species, declare the Vision of healing a reality, and, focusing all our efforts toward this purpose, we make inalienable the right and responsibility to nurture and sustain the HEART OF HUMANITY. Like the Council of the Grandmothers so conveniently ignored by our Founding Fathers as the fourth part of the balance necessary to sustain a true democracy (see

52

The Iroquois Confederacy), we must care not only
for this generation but also consider the needs of
the fourth to the seventh generation to come.
□Anyone who doesn't believe that the necessary
steps must be taken now to ensure the survival of
the planet, with Clean air, water and soils a
necessity, not a luxury of the rich and famous, is
an alien and should be sent back to whatever
previous galaxy or planet they came from to enjoy
there the fruits of their labor, worlds without
habitat, species or diversity: worlds now probably
incapable of sustaining life itself, the most
miraculous and precious of gifts. □Successful and
nurturing ecospheric conditions are created by the
concept of "Heaven's Flame"--small groups of people
living together and working in harmony to support
themselves: i.e. ecological, atmospheric,
earthworm-approved agriculture creating habitat
that nourishes and improves conditions for
conscious living and interaction with the earth as

a biosphere capable of regeneration with "husbandmanship" or conscious engineering of habitat or permaculture in it's many forms and diverse mechanisms allowing abundance without pollution. This harmony presupposes the need and the desire to form "self-activated group consciousness" (the next evolutionary step?), allowing a global mechanism in the earth's body to supersede the need for competition in order to form a more perfect union, like the organism of the human body itself, where all effective functioning is directed toward the survival of the globe as a life-sustaining organism, treating all polluters as cancerous cells that must be stopped immediately before the very entity the parasitical, life-threatening corporations feed on be reduced to a mars-like condition no longer capable of supporting life as we know and enjoy it now. ☐

Appendix D: **N-DIMENSIONAL PROXIMITY APPROACH**

THE N☐DIMENSIONAL KNOWLEDGE PROXIMITY APPROACH TO TECHNOLOGY ASSESSMENT: THE CASE OF QUANTUM ELECTROMAGNETIC SYSTEMS

Fred Bernard Wood, Sr., Ph.D. (Electrical Engineering) P.E.
Computer Social Impact Research Institute, Inc. (csiri.org)
(deceased, 2006).

1. Abstract

Forecasting and assessing the development of revolutionary technologies pose a substantial challenge because the techniques such as trend extrapolation, evolutionary projections, substitution curves, cybernetic systems analyses, and the like are limited in their ability to capture "out of the box" or "wild card" possibilities. Over the last several decades, successive iterations have been developed of a multi☐dimensional systems approach to assessing revolutionary technologies (Wood, 1979, 1996). This approach includes a structure for relating knowledge about classes of activity, levels of phenomena, and stages of evolution. It is called the "N☐Dimensional Knowledge Proximity" systems approach, because it emphasizes an identification of almost current knowledge up to 1992 (six years old to cooperate with military institutions to protect security for six years) about a technology so as to identify and understand the linkages between different dimensions or variables, and to provide a robust basis for projecting future developments and implications. In this paper, the history and current status of the Knowledge Proximity Approach is summarized, with emphasis on refinements that have been made since the 1996 ISSS meeting in Budapest, Hungary. The objective in limiting security classification of new weapons concepts to six years is to let religious and public service organizations have a general knowledge of the trend in weapons development from 1946 to 1992 so that they can start work on the ethical problems involved with new weapons systems without disturbing the secrecy of the latest weapons development. In World War II the United States tried to keep the Atomic bomb secret until after we had killed 66,000 people and wounded 69,000 people at Hiroshima on Aug. 6, 1945. This did not give the people who specialized in dealing with ethical questions a chance to sort through the ethical questions until after the new weapon had been used. A more humanitarian way would be to have a United Nations Committee discussing the ethics of the new class of weapons before some conflict escalates into World War III. The National Council of Churches, the Buddhist Peace Fellowship, and other groups could discuss with military leaders whether six years is adequate or whether more time like ten years is needed by the military institutions for initial new weapons design.

Keywords: Creation, Electromagnetics, Ethics, Weapons, Medicine
WP8S1724

See this paper online at:

ABOUT CSIRI:

CSIRI The Computer Social Impact Research Institute (CSIRI)

was founded in 1978 as a vehicle for assessing the social impacts

of computer technologies. •The star of our series was

Dr. Fred B. Wood, III, PhD, EE• President and Founder of CSIRI:

•COMPUTER SOCIAL IMPACT RESEARCH

INSTITUTE, INC. •"Computers for peace, freedom and prosperity." ••CSIRI was first incorporated in 1979 in California as a non-profit corporation. However, Dr. Wood has hardly ever used his status as a federally-approved non-profit to do any fundraising, but has only used his own resources in his research.••

CSIRI is dedicated to research and education on the social impacts of computing,• including the Internet, and related science and technology.••

CSIRI provides specialized "StarShip Earth" research •and consulting services on the impacts of quantum electromagnetics •on energy, transportation,• environment, social equity and civil liberties,• global diplomacy and peaceful coexistence, national security and related areas.

CSIRI uses an holistic systems science methodology to analyze the complex interactions between computers and society.

The CSIRI Founder and President, Dr. Frederick B. Wood, III, •held the PhD, MSEE and BSEE degrees from the University of California at Berkeley,• worked on microwave radar applications at MIT (Massachusetts Institute of Technology)•and Berkeley

Radiation Laboratories and served for 35 years on the technical

staff of IBM Corporation.•

The CSIRI President was also a founding Member of the

International Society for the Systems Sciences (ISSS),

(formerly known as the Society for General Systems Research.)•

He was an honored lecturer, last presenting a paper on Climate

Change and Electromagnetics to ISSS in Toronto in July, 2000.

S•Dr. Wood is generally known as the "Father of General Systems

Theory."•

CSIRI has been re-founded in WA in 2007 by Andrea Bowe,

Dr. Wood's research assistant, who has

 a Design Team with "win-win" solutions to Earth Regeneration.

Dr. Wood participated in several Global Crisis Solutions

Conferences since 1998.•

GCSC 2 was also shown in July 1999 on communitytv.org

Produced by Andi Bowe.

 •Dr. Wood is in

America's Registry of Outstanding Professionals 2003.•

Here is a link to this history online:

http://www.angelfire.com/hi3/spiritualun/csiri.html
or

http://www.csiri.org

Email abowe@csiri.org for further updates.

Lulu.com also has a text for download or print by Andrea Bowe

On Quantum Physics and Musical Functionality.

BIBLIOGRAPHY

Anderson, Carolyn, editor. (1993). *The Rings of Empowerment*. San Anselmo, CA.: Global ⯐Family

Bearden, Tom. (1986). *Fer-De-Lance*. Chula Vista, CA: Tesla Publishing Co. http://www.cheniere.com Bearden's page soliton@bellsouth.net

Becker, Robert O. (1985). *The BODY ELECTRIC: ELECTROMAGNETISM AND THE FOUNDATION OF LIFE*. NY: Morrow.

Begich, Dr. Nick and Roderick, James. (2000). *Earth Rising: the Revolution Toward a Thousand Years of Peace.* Anchorage, Alaska: Earthpulse Press.

Bohm, David. (1995). *UNFOLDING MEANING: A Weekend of Dialogue.* 29 W. 35thSt. NY, NY 10001: Routledge, ISBN 0 415 13638 5

Brennan, Ann Barbara. (1987). *Hands of light: A Guide to Healing Through the Human Energy. And (1993) LIGHT EMERGING: The Journey of Personal Healing.* NY: Bantam Books.

Childre, Doc and Howard Martin with Donna Beech. (1999). *THE HEARTMATH SOLUTION.* SF, CA: Harper. .

Childress, David Hatcher, compiler. (1994). *THE FREE-ENERGY DEVICE HANDBOOK: A Compilation of Patents and Report.* Illinois: Adventures Unlimited Press.

Clow, Barbara Hand. (1995). *The Pleiadian Agenda: A New Cosmology for the Age of Light.* Santa Fe, NM:

Bear and Company, Pub.

Consoletti, N. (1998) Ph.D. Thesis: "Theory of
Whole Systems". ncckc@pocketmail.com

Fagg, Lawrence W. (1999). *Electromagnetism and the
Sacred* . NY: Continuum.

Felix, Robert W. (1997). *Not By Fire but by Ice.*
Bellevue, WA: Sugarhouse Publishing.

Gordon, Richard. (1999). *Quantum Touch: The Power
to Heal.* Berkeley, CA: North Atlantic Books.

Hamaker, John D., and Weaver, Donald A. (1982). *The
Survival of Civilization.* Burlingame, CA. Hamaker-
Weaver Publishers.

Hamaker, John D., and Weaver, Donald A (1986). "The
Climate Cycle", An Extract from □"The Hamaker
Thesis on Survival" SGSR 30th Annual Meeting, Vol.
II. L-31 to L-44. University of Pennsylvania.

Henderson, Hazel. (1996). *Building a Win-Win World: Life Beyond Global Economic Warfare.*SF: Berrett-Koehler Pub.

Hock, Dee. (1999). *Birth of the Chaordic Age.* SF: Berrett-Koehler Pub.

Hubbard, Barbara Marx. (1998). *Conscious Evolution: Awakening the Power of Our Social Potential* and (1995) *The Revelation: A Message of Hope for the New Millenium.* Novato, CA: New World Library.

Hunt, Bruce J. (1991). *The Maxwellians.* NY: Cornell Univ. Press

Hurtak, J.J. (1977). *YHVH The Book of Knowledge: The Keys of Enoch.* Los Gatos, CA: The Academy for Future Science. 76-55939

Johari, Harish. (1987).*CHAKRAS.* Rochester, Vermont: Destiny Books.

Karl, Thomas R., and Trenberth, Kevin E. (1999).

"The Human Impact on Climate." Scientific American. Nov./Dec.

Keller, C. F. (June 1998). "Global Warming: An Update" Los Alamos, NM:Institute of Geophysics and Planetary physics. http://www. igpp.lanl.gov/ IGPP

Kessler, Ronald. (1995) *Inside the White House.* NY: Pocket Books.

Kraft, R. Wayne. (1983). *A Reason To Hope: A Synthesis of Teilhard de Chardin's Vision and Systems Thinking.* Seaside, CA: Intersystems Publications.

Kukla, Geroge and Went, Ellen. (1992). *Start of a Glacial.* Berlin: Springer-Verlag.

Laszlo, Ervin, Grof, Stanislav and Russell, Peter. (1999). *A Transatlantic Dialogue.* Boston, Mass.: Element.

LaViolette, Paul A. (1994). *SUBQUANTUM KINETICS The*

Alchemy of Creation 1176 Hedgewood Ln., Schenectady, NY 12309: ISBN 0 9642025 0 6 And (1995) Beyond the Big Bang. Vermont: Park Street Press.

Manning, Jeane. (1996). *The Coming Energy Revolution: The Search for Free Energy.* Garden City Park, NY: Avery Publishing Group. ISBN:0-89529-713-2

Ornstein, Robert. (1997). *The Right Mind: Making Sense of the Hemispheres* .NY: Harcourt, Brace and Co.

Peat, F. David. (1997). *INFINITE POTENTIAL: The Life and Times of DAVID BOHM.* Menlo Park, CA: Addison-Wesley.

Pond, Dale, and Keely, John, and Tesla, Nikola and others. (1990, 1996) *UNIVERSAL LAWS NEVER BEFORE REVEALED: KEELEY'S SECRETS Understanding and Using the Science of Sympathetic Vibration.* Santa Fe, NM:

The Message company.

Quigg, Chris. (1983,1997) *GAUGE THEORIES OF THE STRONG, WEAK, AND ELECTROMAGNETIC INTERACTIONS.* Reading, Mass: Adison Wesley Longman, Inc. □ISBN 0-201-32832-1.

Schneider, Michael S. (1994). *A Beginner's Guide to Constructing the Universe: The Mathematical Archetypes of Nature, Art and Science.* NY: HarperPerennial.

Schneider, Stephen H. (1989) *,GLOBAL WARMING ARE WE ENTERING THE GREENHOUSE CENTURY?* SF: Sierra Club Books.

Sharpe, Kevin J. (1993). *David Bohm's World New Physics and New Religion.* 440 Forsgate Dr. Cranbury, NJ, 08512:Associated University Presses.

Sitchin, Zecharia. (1985). *The Wars of God and Men.* NY: Avon books. See also *The 12th Planet* and *The Earth Chronicles* by this author at avonbooks.com.

Starhawk, (1993). *The Fifth Sacred Thing.* NY, NY:
Bantam Books.

Swartz, Tim. (1998). *Secret Black Projects of the
New World Order: Anti-Gravity UFO's, □Black
Helicopters and Mysterious Flying Triangles*. New
Brunswick, NJ: Abelard Productions □Publishing.

Targ, Russell and Katra, Jane, Ph.D. (1998).
*Miracles of Mind: Exploring Nonlocal □Consciousness
and Spiritual Healing.* Novato, CA.: New World
Library.

Tompkins, Peter, and Bird, Christopher. (1989).
*SECRETS OF THE SOILS New Age Solutions □for
restoring our planet*. NY: Harper and Row. ISBN 0 06
091968 X

Walker, Evan Harris. (2000). *The physics of
CONSCIOUSNESS*. Cambridge, Mass.: Perseus □Books.

Weinberg, Stephen. (1996). *THE QUANTUM THEORY OF
FIELDS Vol. II Modern □Applications*. ISBN 0 521

58555 4, 40 W. 20th St. NY NY 10011-4211: Cambridge
University Press

Wood, Fred B., Sr. (1998). "*The N-Dimensional
Knowledge Proximity Approach To Technology
Assessment: The Case of Electromagnetic Systems.*
ISSS Proceedings. csiri.org

Wood, Fred B., Sr. (1992)." *Climate Cycles and the
Development of Civilization"* Whose World to Lose?
Berkeley, CA.: Earth Regeneration Society.

Wood, Matthew. (1987). *Seven Herbs: Plants as
Teachers.* Berkeley, CA: North Atlantic Books.

Yogananda, Paramahansa. (1946). *Autobiography of a
Yogi* .LA, CA: Self-Realization Fellowship.

ISBN: 978-0-557-03024-8

Published by lulu.com

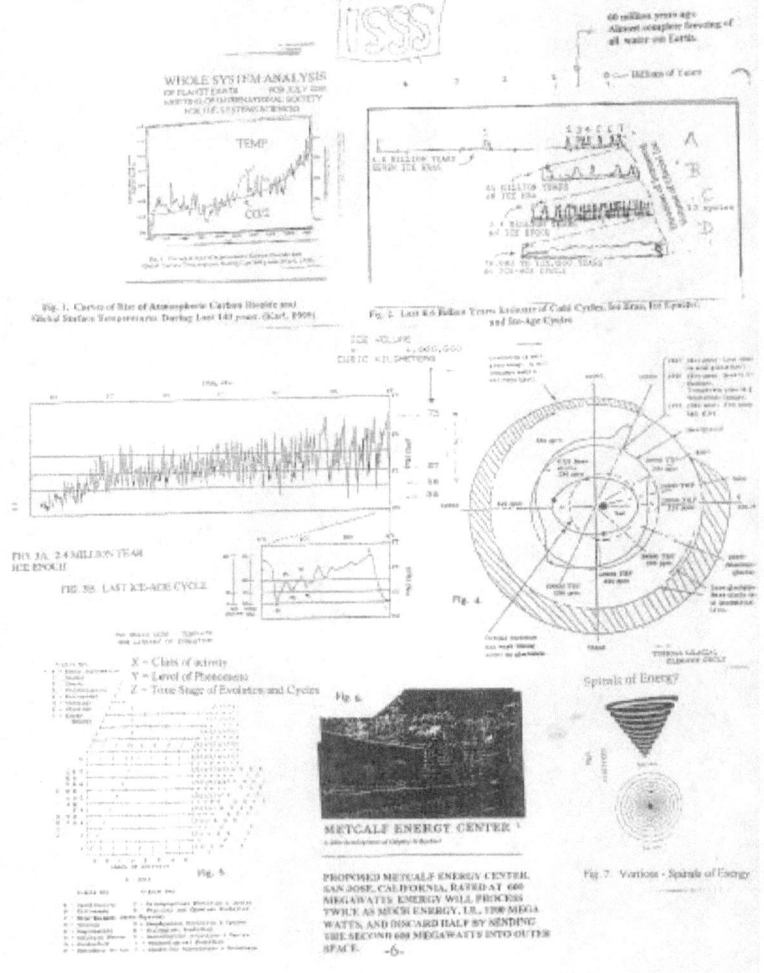

WHOLE SYSTEM ANALYSIS
OF PLANET EARTH
FOR JULY 2000
MEETING OF INTERNATIONAL SOCIETY
FOR THE SYSTEM SCIENCES

Fig. 1. Curves of Rise of Atmospheric Carbon Dioxide and
Global Surface Temperatures During Last 140 years (Karl, 1999)

Fig. 2. Last 4.6 Billion Years Inventory of Cold Cycles, Ice Eras, Ice Epochs,
and Ice-Age Cycles

FIG. 3A. 2.4 MILLION YEAR
ICE EPOCH

FIG. 3B. LAST ICE-AGE CYCLE

Fig. 4.

X = Class of activity
Y = Level of Phenomena
Z = Time Stage of Evolution and Cycles

Fig. 6.

Spirals of Energy?

Fig. 5.

METCALF ENERGY CENTER

PROPOSED METCALF ENERGY CENTER,
SAN JOSE, CALIFORNIA, RATED AT 600
MEGAWATTS. ENERGY WILL PROCESS
TWICE AS MUCH ENERGY, I.E., 1200 MEGA
WATTS, AND DISCARD HALF BY SENDING
THE SECOND 600 MEGAWATTS INTO OUTER
SPACE.

Fig. 7. Vortices - Spirals of Energy

-6-

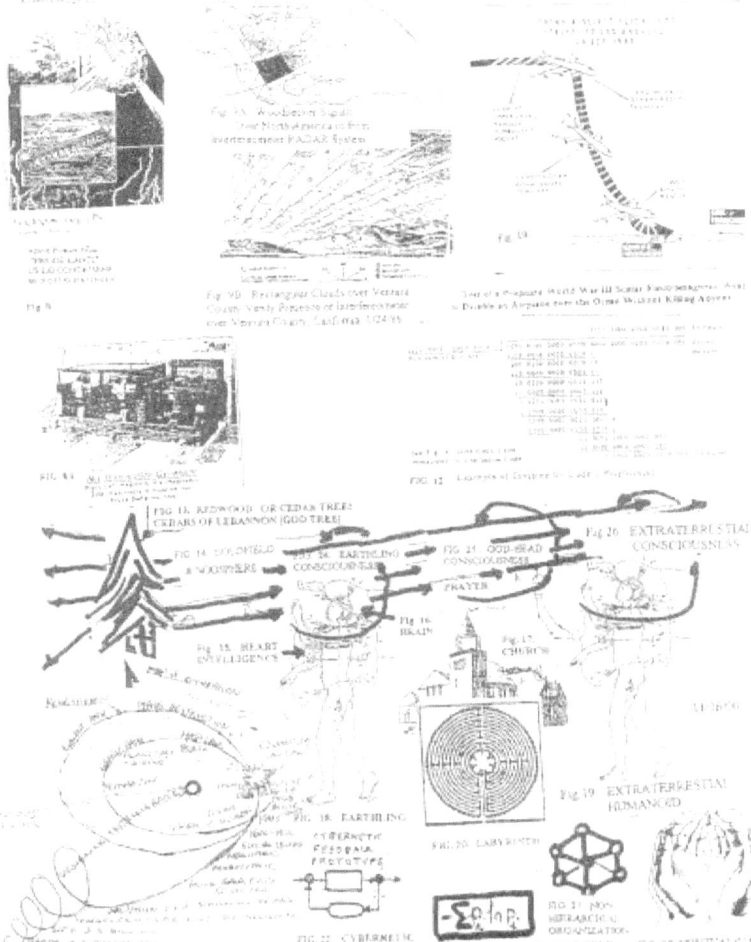

Fig. 13. Woodpecker Signal over North America from over-the-horizon RADAR System.

Fig. 9b. Rectangular Clouds over Ventura County. Newly Detected Interferometer over Ventura County, California 1/24/95

Fig. 9

Test of a Proposed World War III Scalar Electromagnetic Field to Disable an Airplane over the Ocean Without Killing Anyone

FIG. 12

FIG. 13. REDWOOD OR CEDAR TREES CEDARS OF LEBANNON (GOD TREE)

FIG. 14. BIOFIELD & WOOSPHERE

FIG. 24. EARTHLING CONSCIOUSNESS

Fig 25. GOD-HEAD CONSCIOUSNESS

Fig 26. EXTRATERRESTIAL CONSCIOUSNESS

PRAYER

Fig 16. BRAIN

Fig. 17. CHURCH

FIG. 15. HEART INTELLIGENCE

FIG. 18. EARTHLING CYBERNETIC FEEDBACK PROTOTYPE

FIG. 19. EXTRATERRESTIAL HUMANOID

FIG. 20. LABYRINTH

FIG. 21. SPIRAL OF EVOLUTION

FIG. 22. CYBERNETIC FEEDBACK

FIG. 23. HEART AND HOLOFIELD NEGENTROPY PRODUCTION

FIG. 21. NON HIERARCHICAL ORGANIZATION

FIG. 28. SPIRITUAL GOD UNDERSTANDING

www.ingramcontent.com/pod-product-compliance
Lightning Source LLC
Chambersburg PA
CBHW021023180526
45163CB00005B/2083